Pluto's orbit is billions of miles from the Sun

Pluto

Steve Potts

Smart Apple Media

Comsewogue Public Library
170 Terryville Road
Port Jefferson Station, NY 11776

OUR SOLAR SYSTEM

☀ Published by Smart Apple Media

1980 Lookout Drive, North Mankato, MN 56003

Designed by Rita Marshall

Copyright © 2003 Smart Apple Media. International copyright reserved in all countries. No part of this book may be reproduced in any form without written permission from the publisher.

Printed in the United States of America

☀ Pictures by Lowell Observatory, NASA, Photo Researchers (Chris Butler/Science Photo Library, Lynette Cook/Science Photo Library, Mark Garlick/Science Photo Library, Science Photo Library, Space Telescope/Science Institute/Science Photo Library), Tom Stack & Associates (Brian Parker, TSADO/NASA)

☀ Library of Congress Cataloging-in-Publication Data

Potts, Steve. Pluto / by Steve Potts. p. cm. — (Our solar system)

Includes bibliographical references and index.

☀ ISBN 1-58340-101-6

1. Pluto (Planet)—Juvenile literature. [1. Pluto (Planet)] I. Title.

QB701 .P68 2001   523.48'2—dc21   2001020120

☀ First Edition 9 8 7 6 5 4 3 2 1

Pluto

CONTENTS

The Smallest Planet — 6

Pluto's Surface — 8

Pluto's Moon — 14

Exploring Pluto — 16

Additional Information — 24

The Smallest Planet

To the ancient Romans, Pluto was the god of the dead, the keeper of the underworld. He ruled a dark wasteland. The darkest, most desolate planet in our solar system is also called Pluto. This tiny planet, almost six billion miles (9.6 billion km) from the Sun and smaller than Earth's moon, is the furthest planet in our solar system. It is so distant and small that it was not discovered until the 1930s. Pluto is still hard to see, even through many modern **telescopes**. Most planets

Frozen gases give Pluto a pinkish color

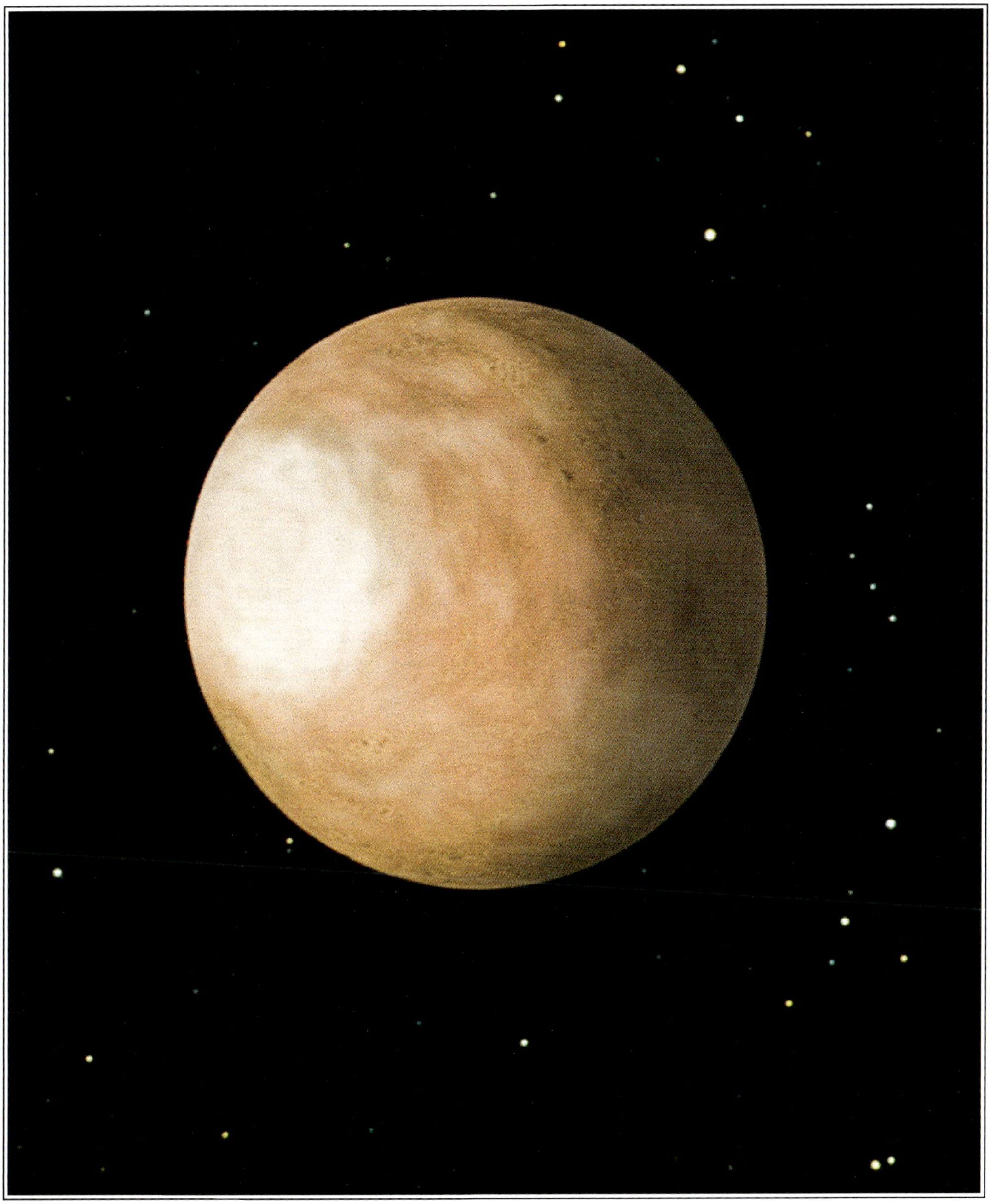

orbit the Sun in a rough circular pattern, but not Pluto. Its orbit follows an extreme oval course. Sometimes Pluto's orbit brings it to within 3.7 billion miles (5.9 billion km) of the Sun. At other points in its orbit, Pluto is 7.3 billion miles (11.8 billion km) from the Sun.

Pluto's Surface

Pluto is a cold, dark place. Scientists believe that it has a solid center of rock that is covered by a layer of water and ice. It is also possible that Pluto's surface is covered with the gases known as methane, ethane, and carbon monoxide. This mixture of gases would be frozen into ice. ❄ Since Pluto receives little light from the distant Sun, its surface temperatures usually do not reach above −364 °F (−220 °C). Some views of

Scientists think Pluto has a rocky center covered by ice

Jupiter

Sun

Mercury

Artwork showing how the planets and Sun compare in size

Pluto / Charon

Saturn

Uranus

Neptune

Venus

Earth / Moon

Mars

Pluto through telescopes hint that it has "warm" and "cold" areas on its surface. The "warm" places show up as darker areas in photographs. Even the "warm" areas are cold, though. They are probably about −328 °F (−200 °C).

☀ Earth's **atmosphere** is mostly nitrogen and oxygen, but on Pluto, the air is mostly nitrogen and methane. This combination of gases would be deadly for humans. The atmosphere probably also changes as Pluto approaches the Sun in its orbit. When Pluto is far from the Sun, the gases in the atmosphere are

Pluto was discovered in 1930 by American astronomer Clyde Tombaugh.

frozen solid. When Pluto gets closer to the Sun, some of the ice

melts and turns back into gases.

Clyde Tombaugh discovered Pluto after years of searching

Pluto's Moon

In 1978, astronomer Jim Christy discovered another fascinating fact about Pluto. The planet, which is only one-fifth the size of Earth, has a moon half the size of Pluto. In fact, its moon is so big and Pluto so small that some astronomers call Pluto a "double planet." Pluto's moon was named Charon (pronounced "Karen") after the mythical figure who led the dead to the god Pluto's underworld. Located about 12,400 miles (20,000 km) from Pluto, Charon seems to be made up

Pluto has just two-tenths of one percent of Earth's mass.

mostly of ice with a rocky center. Scientists believe that

Charon's dark surface means that its temperature is a little

higher than Pluto's. ✷ Some astronomers believe that Pluto

Pluto and Charon may have once been a single moon

and Charon were once a single moon of Neptune. Their theory suggests that this moon was struck by something very large, which broke the moon in two and tossed the pieces out into space.

Exploring Pluto

Most of what has recently been learned about this tiny planet has come from the *Hubble Space Telescope*. The *Hubble* took a picture of Pluto and its moon when Pluto's orbit brought it within 2.6 billion miles (4.2 billion km) of Earth. Even though the planet looked like a fuzzy soccer ball in

An illustration of the powerful *Hubble Space Telescope*

the picture, the *Hubble* did show Pluto and Charon as two clearly separate disks. This allowed astronomers to accurately measure the sizes of Pluto and its moon. Pluto is 1,457 miles (2,345 km) wide, and Charon is 790 miles (1,270 km) wide. It was also discovered that Charon is more blue in color than Pluto. ☀ Even with the help of the *Hubble*, scientists have not learned many of Pluto's secrets yet. The planet is simply too far away for any Earth-bound telescope to get a good picture of it. No probes have gone to Pluto yet, but NASA plans to send a pair of probes, called the *Pluto-Kuiper Express*, past Pluto in December 2012. ☀ Probes are small spacecraft that are sent into space by rockets or are carried on a space shuttle.

Before Pluto's existence was confirmed, astronomers called it Planet X.

The *Pluto-Kuiper Express* probes may one day visit Pluto

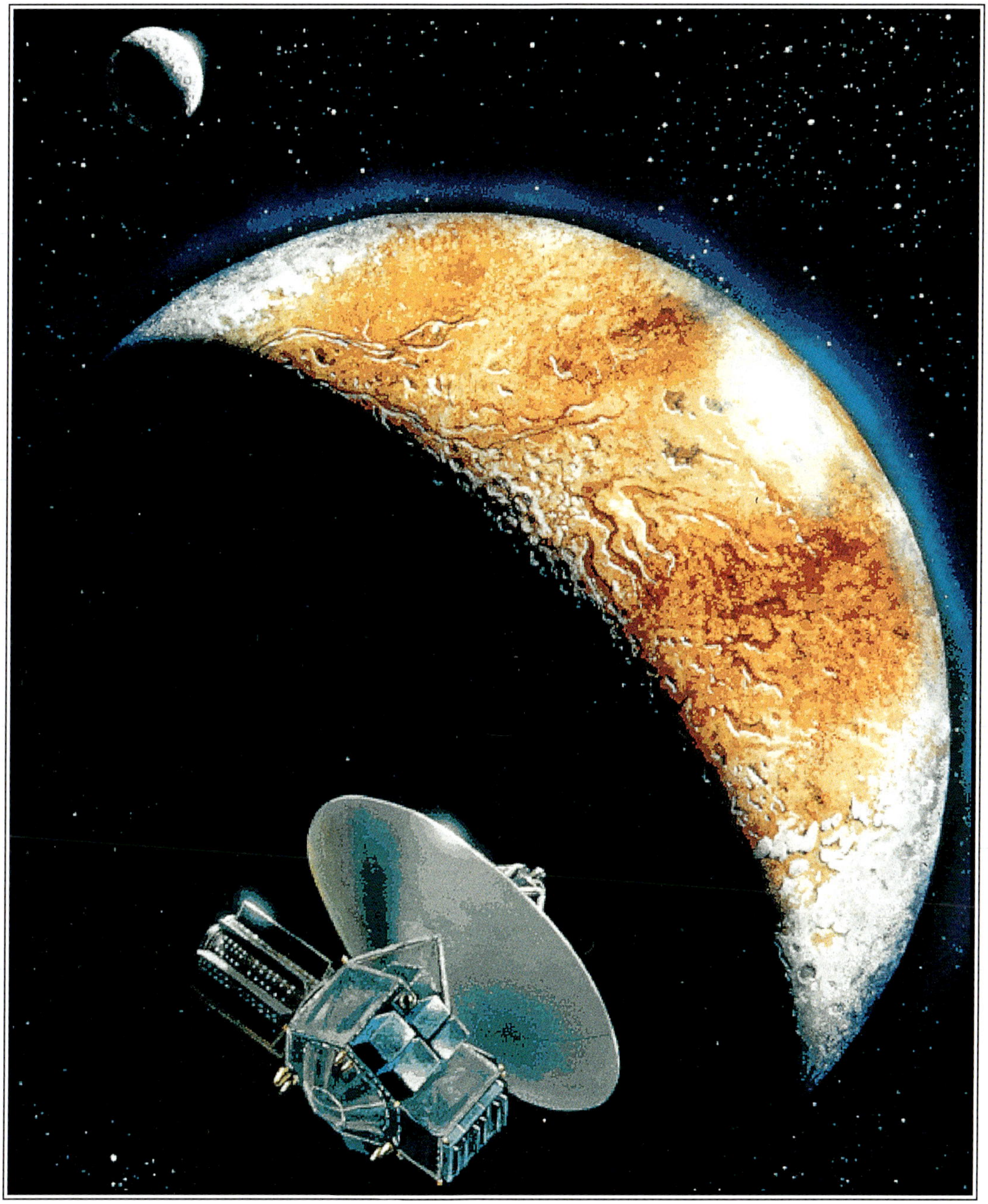

These probes carry computer and photographic equipment.

The cameras take pictures that are sent back to Earth by **radio waves**. ☀ The *Pluto-Kuiper Express* will take about 12 years to reach Pluto. They will photograph Pluto, then head further into outer space. Scientists hope that these photographs will help us learn more about this strange, small planet.

Large antennas collect radio waves sent through space

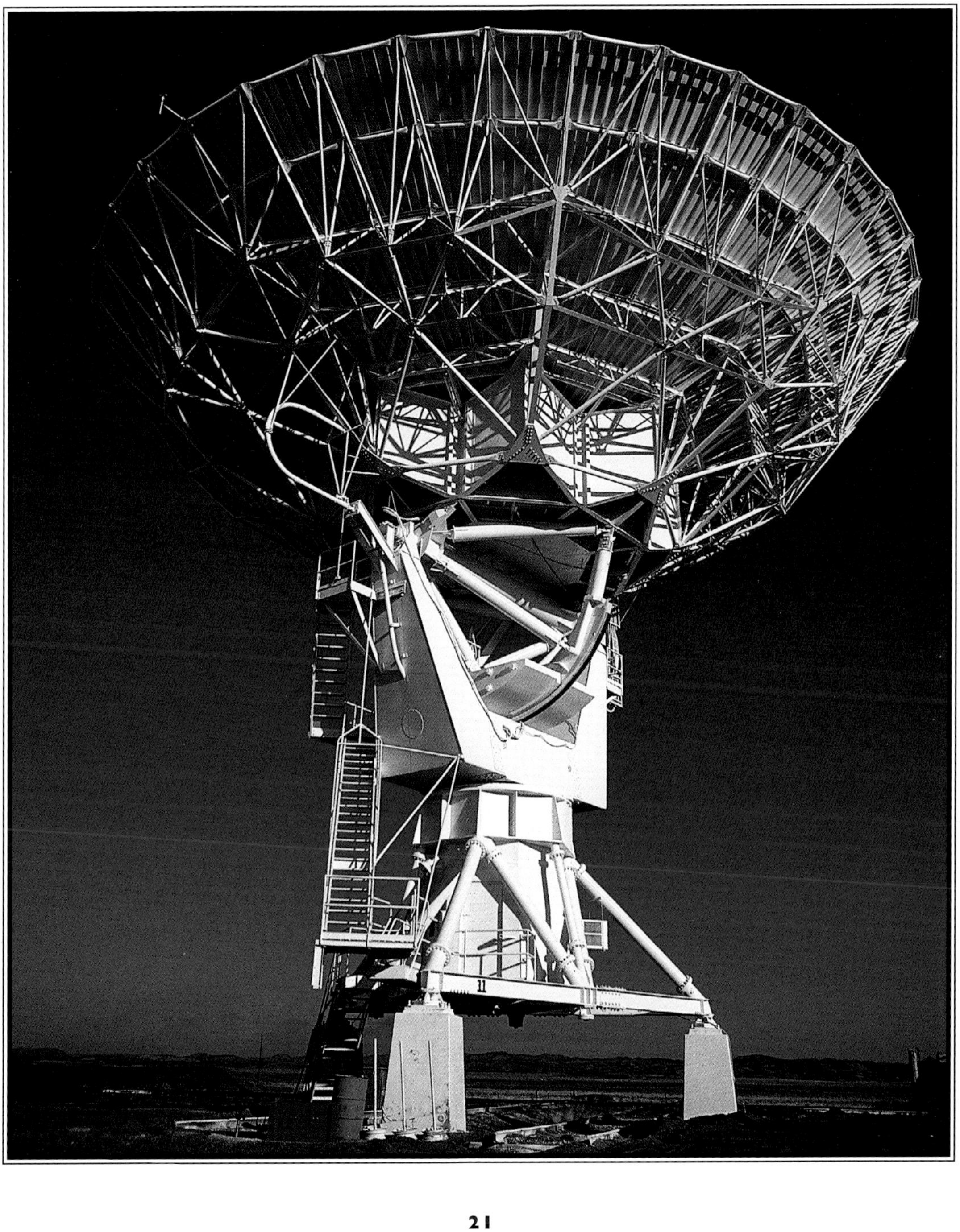

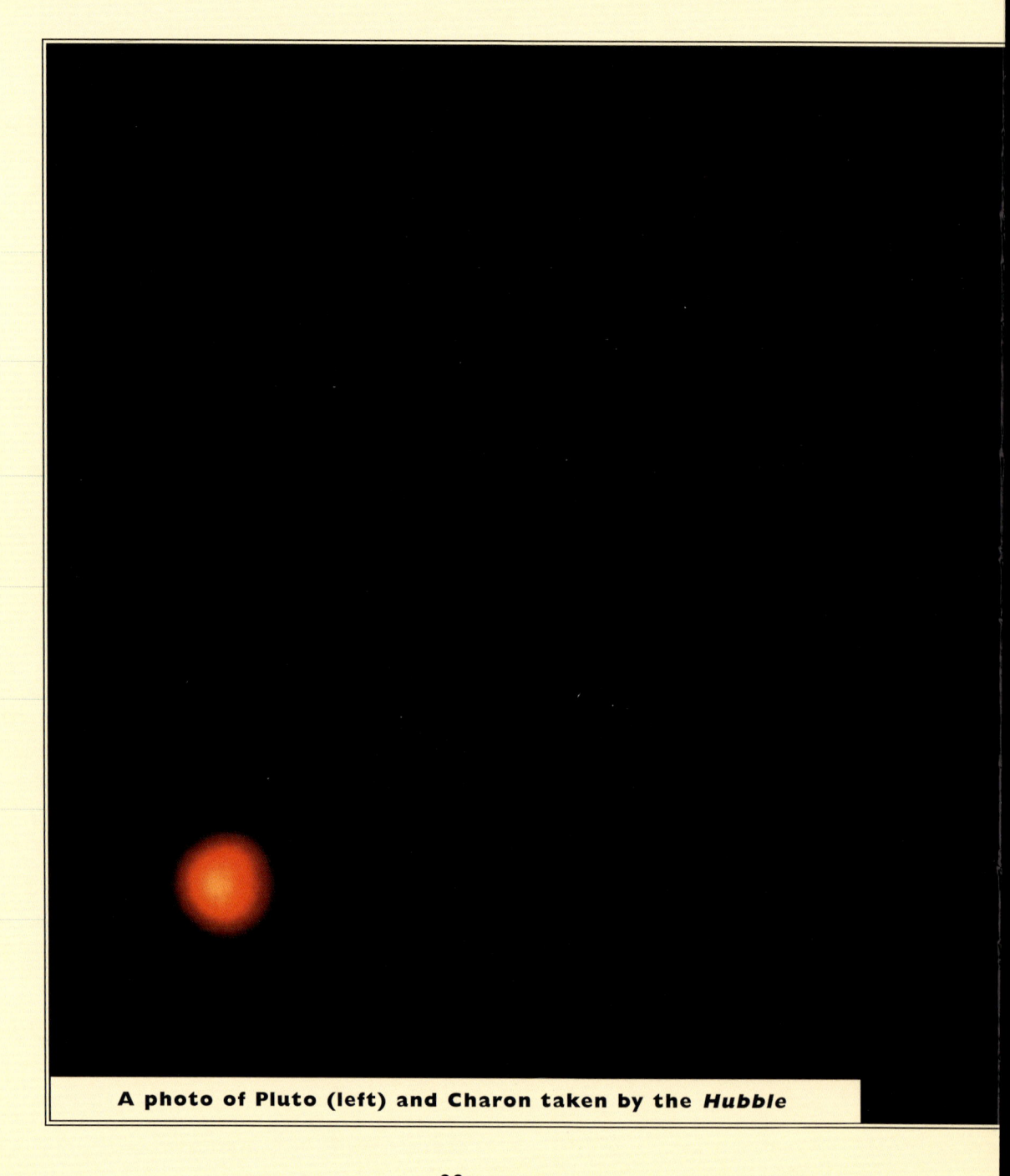

A photo of Pluto (left) and Charon taken by the *Hubble*

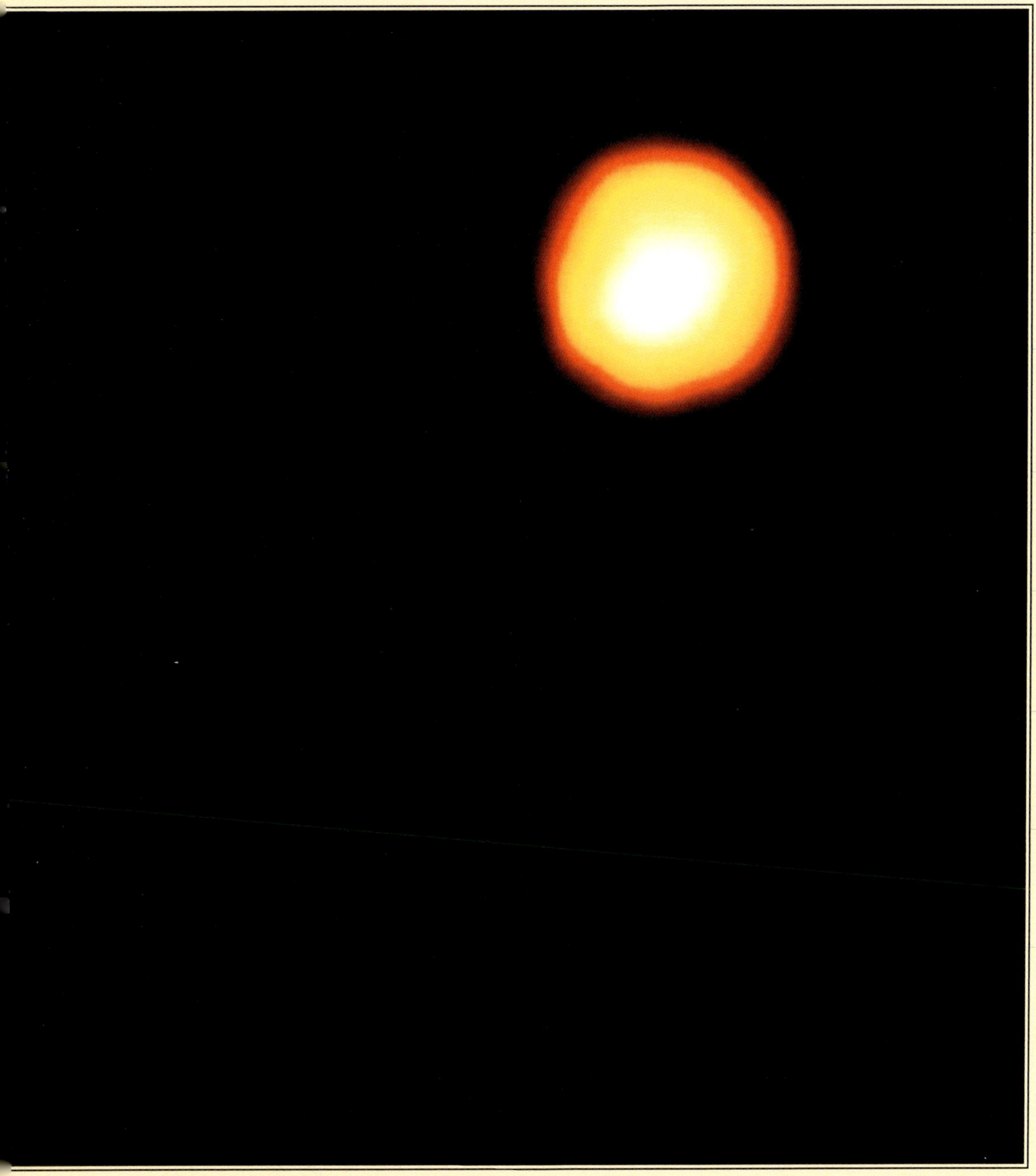

Index

atmosphere 12-13
Charon 14-16, 17
Hubble Space Telescope 16-17, 18
orbit 8, 12-13, 16
probes 18, 20
size 14, 17
surface 8-9, 12
temperatures 9, 12, 15

Words to Know

atmosphere—the nearly invisible layer of gases that surrounds a planet

orbit—travel in a repeating circular pattern around another object

radio waves—energy that travels at the speed of light (186,000 miles/second or 300,000 km/sec) from one antenna to another to form a message

telescopes—instruments that use glass lenses to magnify distant objects

Read More

Asimov, Isaac. *A Double Planet? Pluto and Charon*. Revised and updated edition. Milwaukee, Wisc.: Gareth Stevens Publishing, 1996.

Bond, Peter. *DK Guide to Space*. New York: DK Publishing, 1999.

Couper, Heather, and Nigel Henbest. *DK Space Encyclopedia*. New York: DK Publishing, 1999.

Furniss, Tim. *Atlas of Space Exploration*. Milwaukee, Wisc.: Gareth Stevens Publishing, 2000.

Internet Sites

Astronomy.com
http://www.astronomy.com/home.asp

NASA: Just for Kids
http://www.nasa.gov/kids.html

The Nine Planets
http://seds.lpl.arizona.edu/nineplanets/nineplanets

Windows to the Universe
http://windows.engin.umich.edu

COMSEWOGUE PUBLIC LIBRARY
170 TERRYVILLE ROAD
PORT JEFFERSON STATION
NEW YORK 11776